HACKING THE EARTH
UNDERSTANDING THE CONSEQUENCES
OF GEOENGINEERING

by
Jamais Cascio

Contents

PREFACE

This book collects essays I've written over the past five years about the subject of "geoengineering," both on the environmental blog I co-founded, Worldchanging.com, and my solo futurism blog, OpenTheFuture.com. Most of the pieces have been modified slightly to make a more coherent document, and to fix the various bloggy references to articles "published today" and items "posted last week."

The subject of geoengineering is both a special interest of mine and an increasingly important matter of public and scientific debate. It's something of a staggering concept: intentionally altering geophysical systems in order to bring about desired changes to the global climate. Can we really do this? If so, what would it mean? This book focuses on the latter question, and is written for people who have a reasonably solid grasp of how global warming works, and what it will take to stop it.

This book is also an experiment in self-publishing. I've wanted to write a book (other than my game setting books, *Transhuman Space: Broken Dreams* and *Transhuman Space: Toxic Memes*, published by Steve Jackson Games) for some time, and in recent weeks various friends have tried out — and given high marks to — web-based self-publishing outfits like Lulu.com. While I still would like to produce a book the traditional way, I thought I'd give this method a shot.

INTRODUCTION:
SOLVING THE CLIMATE CRISIS

The debate about global warming is over. While the remaining holdouts and dead-enders continue to bray about hoaxes and imaginary disputes, the dominant focus of conversation about climate disruption for serious people boils down to a simple question: *what do we do about it?*

A simple question, but not a simple answer, in part because there are multiple possible responses, and they're not necessarily mutually-compatible. They cover three broad categories: **Prevention** (actions that reduce the risks of global warming or soften its eventual impact); **Mitigation** (actions directed at reducing the harm of global warming, and as possible reducing its sources); and **Remediation** (actions intended to reverse global warming and its effects). Each of these entails its own set of political, economic and environmental risks.

---

* *Original version of this piece published at Open the Future in October of 2007.*

One of the reasons why the answers are not cut-and-dried is an aspect of global warming that, as yet, still does not receive the kind of mainstream attention it deserves: *climate commitments*. It turns out that, no matter what we do, we are committed to a certain amount of continued warming and climate change. Moreover, the longer we wait to start acting seriously, the more of a commitment we'll build up.

Much of this commitment comes from the physics of climate change. There is an enormous amount of lag in geophysical systems. We see that in particular in the delay between actions that increase or decrease climate forcings and the resulting climate impacts. Some of that lag comes from how long it takes for certain chemicals to cycle out of the atmosphere, some comes from how warming itself changes natural cycles, and much of it comes from the thermal inertia of the oceans — the slow pace at which ocean temperatures change. Climate scientists generally describe this climate lag as being around 20-30 years; even if we were to cut off all additional carbon emissions right this very second, we'd still see another two to three decades of warming.

That's if we're lucky. If, in that 20-30 years, the rising temperatures start triggering climate feedback effects (such as large-scale methane emissions from melting permafrost, or the reduction of the polar ice cap causing more heat to be absorbed by the dark water), problems could continue even past the 20-30 year mark. And, of course, we're not going to cut off all additional carbon emissions any time soon, so that 20-30 year countdown hasn't even started.

It should be clear at this point that the longer we wait, and the more of a climate commitment we build up, the more likely it is that we'll see feedback effects.

So with that, here are the three key solution arguments for climate disruption:

PREVENTION

The potential for dangerous feedback effects and other disasters forms the key driver for the *Prevention* argument. Given that we simply don't yet know how damaging to our environment and our civilization these feedback effects could be, wisdom dictates that we do all we can to start eliminating the anthropogenic sources of greenhouse gases, in order to hold the committed warming to the lowest possible amount.

There are two leading versions of the Prevention argument.

The first is that, while no one single solution will solve all aspects of the global warming crisis, we have at our fingertips a sufficient variety of partial solutions that, in combination, would be able to reduce and eliminate greenhouse gas emissions in a short enough time-frame to avoid the worst of the climate disruption threats. This version is best captured in Princeton scientist Robert Socolow's "Stabilization Wedges" model, which he first wrote about in 2004, and which later appeared in *An Inconvenient Truth*. The wedge model seeks to combine different carbon-reduction solutions, such as trading coal plants for wind power or capturing the methane emissions from cattle, in order to slow, stop, and then reverse the growth of atmospheric greenhouse gases.

Some of the wedges Socolow proposes are technological, but many would best be described as "behavioral" — changes to how we move around, for example, or how we build our cities. Although they won't necessarily be as shiny and fun as new technology, behavioral changes can be quite powerful. Most of the behavioral changes that advocates of the Prevention strategy suggest are beneficial across multiple problems, for example. Greater emphasis on public transit reduces automobile pollution of all sorts, and helps to ameliorate the impact of peak oil; bicycling and reducing meat intake can vastly improve one's health; local diets and recycling can boost regional economies.

The primary risk of a behavioral model for avoiding climate disruption, conversely, is that, if we assume a realistic transition from current ways of life to the more sustainable versions, it will take a generation or more to make happen. Again, as long as we don't hit a dangerous tipping point in environmental systems, this relatively slow pace could still allow us to avoid global disaster — but there's still tremendous uncertainty around the likelihood and triggers for those tipping point changes.

Another risk of the behavioral approach is the very likely countervailing pressure to avoid making changes. Whether due to cost, convenience, tradition or politics, the social and economic changes necessary to reduce greenhouse emissions will face stiff opposition. These can be overcome, but not without time and effort. And if the necessary behavioral changes have a short-term negative impact on global competitiveness and economic capacity — which is possible, but by no means certain — lagging adopters may have an even greater motive to avoid undertaking the necessary adjustments.

Fortunately, technology changes will help. This is the reason that I suspect that the wedge model would eventually succeed. The wedges Socolow proposes implicitly assume no significant breakthroughs in performance or capabilities over the next fifty year period, an assumption that's hard to justify.

It's also the crux of the second version of the Prevention argument. This version takes the idea of technological change and runs with it, arguing that making the necessary changes now would mean adopting technologies still in flux, when they're overly expensive and likely to be replaced by cheaper, better versions soon. According to this argument, even waiting ten years to start implementing big changes (to infrastructure, transportation, energy production and the like) would mean a significant reduction in overall cost and a likely improvement in how quickly improvement would result.

In a way, this is a variant of the "leapfrog" concept popular in development politics: countries that adopt the current versions of sustainable technology see immediate benefits, but at a high cost and a likely lock-in of particular technologies; countries that wait are able to adopt cheaper, better technologies down the road, and soon overtake the earlier generation nations.

Because proponents of this version of the Prevention argument often use it to dismiss making significant behavioral changes, critics often deride it as a "techno-fix" approach. The bigger flaws, in my view, are that it adds to the overall risk by pushing the climate commitment point further out, assumes the development of radically improved technologies (very likely, but not guaranteed), and — most importantly — sets up a social psychology of wanting to wait for the next big thing before acting. The danger of the leapfrog model is that waiting just a bit longer can mean being able to deploy even better technologies... but by the time those roll out, even better systems would be on the near horizon, making it tempting to continue waiting (and, given the desire by economic and political incumbents to continue to accumulate power and wealth in the existing model, there would be no lack of pressure to do just that).

In the ideal world, behavior + technology gives us the best chance of success at preventing climate disaster, and that the possibility of climate feedback makes it imperative to start making changes as quickly as possible. The likelihood of major technology improvements, however, suggests that our best strategy would be to focus our investments in systems that can be improved relatively quickly and can be replaced relatively easily. In a period of survival pressure, the best evolutionary strategy is iteration and experimentation.

MITIGATION

The second response pathway is that of *Mitigation*. The underlying argument is that climate disaster is already at hand, and we should be able to deal with its results, even as we try to eliminate its sources. This is an explicitly pragmatic approach, acknowledging the likelihood that climate commitments and feedback effects will make our best prevention efforts too little, too late, but recognizing that we still need to reduce the worst of the threats.

The Mitigation argument couples the need for major social, economic, political and technological changes to eliminate greenhouse gas emissions with a focus on making sure that the near-term results of climate disruption can be handled in a way that is ultimately in support of ongoing survival. Because of the split focus, this would likely mean that the Prevention-style changes would come about more slowly, thereby guaranteeing that we'll see climate impacts that we'll need to mitigate. (Of course, Mitigation proponents argue, an emphasis on Prevention doesn't guarantee that these impacts *won't* happen — and then we'll have to engage in catch-up mitigation anyway.) Still, proponents of Mitigation argue that we're better off working in this way *today* than we are waiting until it becomes necessary.

Mitigation efforts could include recognized practices like levees and seawalls, along with a "natural capital" approach that relies on mangroves, marshes and wetlands as storm buffers. It would also likely include developing standards for handling environmental refugees, mechanisms for engaging in climate-disrupted farming, and aggressive deployment of urban agriculture and green roofs. Water rationing, strict limits on energy consumption, and other top-down forms of resource control could be possible elements of a Mitigation model.

In many respects, the behavioral changes required by the Mitigation strategy could be as radical as those in a Prevention approach, and in some cases possibly more so. Mitigation is *not* a best-case response; of all of the different responses discussed here, it's the one most likely to be thought of as a form of "green fascism," especially if the climate disruptions hit faster and harder than initially expected, requiring an equally vigorous response.

Mitigation is also a possible gateway to significant environmental manipulation. We could see focused efforts to alter the genomes of critical plants to make agriculture possible in disrupted ecosystems, or to allow more rapid consumption of CO_2. Such a development would be seen as radical, and could go well beyond what would be considered politically or scientifically acceptable today.

Although it shares many of the same characteristics, the Mitigation argument should not be confused with "adaptation," a buzzword that seems to be popular among the denialists who can't really deny the problem any longer. The crux of the adaptation argument seems to be "lie back and take it:" the disruptions from global warming are too far along to do anything about now, so we may as well do what we can to deal with the results. In many cases, there's an implicit *...and if you can't or can't afford to deal with the results, tough luck*. The adaptation concept emphatically denies the need to make any significant behavioral or radical technological adjustments — what we have works, and trying to impose such changes on a population under pressure just reduces our ability to adapt.

The fallacy at the heart of the adaptation concept is that climate disruption is an either/or deal: either we're in the good present climate, or we're in the bad global warming climate. Unfortunately, climate disruption is a process, not a result. Adaptation without active efforts to prevent worse results simply means having to adapt continuously to increasingly worse and worse environmental effects.

My view is that Mitigation is an increasingly likely approach as we continue to delay serious preventative strategies. If we do see the effects of climate feedback (such as methane excursions or rapid melting of Antarctic and Greenland glaciers), Mitigation strategies become almost over-determined, as — due to climate commitment — there would be simply no way for preventative measures to have a great enough impact swiftly enough to head off disaster.

REMEDIATION

Riskiest of all is the *Remediation* argument. Superficially a mashup of the techno-fix Prevention and the Mitigation strategies, Remediation doesn't look at ways to change greenhouse emissions or deal with direct consequences; rather, its emphasis is on changing the game entirely by altering the core geophysical processes that relate to global warming. Geoengineering — the subject of this book — is at the core of most Remediation concepts, because solving a global-scale problem usually takes a global-scale answer .

Geoengineering is the process of changing the course of climate change by intentionally altering one or more geophysical factors, such as the amount of sunlight hitting the surface of the planet, the overall "reflectivity" of the oceans, or the amount of carbon in the atmosphere. This may be done by boosting existing geophysical or biological processes (such as triggering algae blooms to consume atmospheric CO_2 by fertilizing the ocean), by parallelling natural processes with artificial mimics (such as injecting megatons of sulfate particles into the stratosphere in the manner of volcanic eruptions), or by projects that take advantage of chemistry, but don't echo any normal part of the carbon cycle (such as slow-burning biomass to create "bio-char" in order to bury the carbon underground). Because global warming is, by its very nature, *global*, any attempt at geoengineering would have to be on a vast scale; the effects would need to be felt across large parts of the planet — in some cases, world-wide — in order to have an impact.

The risks associated with geoengineering are correspondingly vast. Although we have a solid understanding of much of the carbon cycle, we're still hazy on many of the complex interactions of geophysical systems. It's possible that we'd get unexpected results, as rapid changes in one system trigger surprising changes in another. It's also possible that we'd find natural feedback processes slowing or even negating our efforts, requiring us to choose between abandoning a project or overdoing it to (hopefully) get the desired result.

These kinds of risks are both widely-acknowledged and very difficult to forecast. Most scientists working on geoengineering ideas recognize the need for keeping a close watch on any such project, in order to pull back if things aren't going as expected. In the section of this book entitled "Geoethics," I go into a bit of detail as to what a philosophy of reversibility might look like.

In many ways, the bigger risk with geoengineering is less-widely acknowledged — and is the primary subject of this book.

As much as we may think about it as a scientific concept, a technological process, or an environmental challenge, geoengineering is also (I would argue primarily) a political and ethical dilemma. Not political in the partisan or left-right sense, but political in the sense that it would be the subject of debates and struggles over power and differing interests between communities, states, and regions. Any attempt to engage in geoengineering will have to deal with questions like:

- *Who decides what the "proper" global climate is?*
- *What happens when one group sees a geoengineering project as a hostile act?*
- *What happens when geoengineering has negative results in some regions, positive in some others? Who's liable?*

All of that said, if preventative and mitigation efforts fail, or if climate scientists determine that climate committment has pushed us to a point where prevention would be insufficient, it is a near-certainty that someone (nation or wealthy non-state actor) will attempt to engage in geoengineering to head off utter disaster, allowing sufficient time for slower preventative solutions to take hold. Unlike preventative measures, which require the active participation of just about everyone to work, geoengineering wouldn't require the cooperation of the global community to carry out. It's worth noting at this point that, although much of the discussion about geoengineering and Remediation has occurred in the United States, the U.S. is by no means the only nation capable of carrying out such an endeavor. Because of that possibility, we need to better understand geoengineering strategies now, so that if the time comes, we'll be able to make smarter decisions about which ones to try and which ones to avoid.

As the scale of the climate emergency continues to grow, the likelihood that we will try some kind of geoengineering also increases. But there's little discussion of geoengineering in the broader public discourse. Some scientists have begun to give the concept deep study, and a handful of journalists have profiled advocates of the idea. Too few voices talk about the political and ethical aspects, unfortunately.

So, can we solve the climate crisis? I think that the evidence is clear that we can, and that we actually have a wealth of potential strategies to do so. We know what kinds of behavioral changes are necessary, and we have a good grasp of the kinds of technologies that would be the most useful. The most important — and uncertain — element, though, is time: the longer we delay, the harder it will be to avoid the worst effects of global warming. We simply can't wait until the big problems start happening — at that point, we'll have committed ourselves to even greater peril over the coming decades, even with a crash preventative effort. This kind of long, slow problem is outside of our common experience, but is increasingly a key characteristic of the challenges we face as a civilization. We can't count on our problem-solving habits to get us out of this one; we need to learn how to integrate foresight and forethought into our policies and everyday lives.

The end result, if we're successful, may be far greater than we dare hope. Not only would we find ourselves in a world of sustainable wealth, abundance and efficiency, we'd be living in a civilization that, for the first time, had really started to think like a mature, adult society.

I'm looking forward to seeing what it's like.

Terraforming Earth:
Understanding Geoengineering

THE GEOENGINEERING OPTION

Here's the *good* scenario: we have maybe a decade, fifteen years on the outside, before we need start seeing a significant and sustained global reduction of greenhouse gases if we are to avoid absolutely catastrophic environmental results. You know the litany by now: unstoppable sea level rise, famine from loss of agricultural land, countless deaths around the world from the heat and opportunistic diseases, extinctions galore, and on and on. If we have the will, ten years would be enough time to implement significant improvements in our transportation and energy technologies, our consumption patterns, and the design of our communities. We know the pieces that we need to put into place, it's just a question of getting them assembled in time.

Probably the best example of what a reasonable greenhouse reduction plan looks would be Robert Socolow's "stabilization wedges" idea, mentioned earlier. With Socolow's plan, no one solution dominates. Well-understood technologies and practices like energy efficiency, renewable energy, biofuels, urban redesign, carbon sequestration, nuclear power, and so forth, all play a role. The stabilization plan takes fifty years to carry out, but starts to reduce greenhouse gases in relatively short order.

But there's a *not-so-good* scenario, too: you know that decade we thought we had? It turns out that it might instead be just a year or two. Good luck.

---

* *Original version of this piece published at Futurismic.com in October, 2006.*

Nearly every plan for dealing with disastrous climate change depends on there being enough time to boost energy efficiency, redesign urban systems, change to cleaner transportation, and so forth, all to reduce greenhouse gas emissions before it's too late. If we don't make it, we're likely to see an environmental transition where the global climate system finds a new stable state quite different from what we have experienced over the history of human civilization — a "tipping point," if you will, into a climate this planet hasn't seen in a million years or more.

According to NASA's Goddard Institute for Space Studies (which specializes in studying the Earth's environmental systems), the Earth may be only a degree (Celsius) or so away from its million year temperature peak, and that degree could mean the difference between a livable climate and a climate tipping point. How long would it take to go up by another degree? We've seen temperature increases of 0.2-0.3 degrees per decade over the past thirty years, so — if NASA is right, and if nothing changes — we'd hit the point of no return (for at least a few tens of thousands of years) by 2040-2050.

But there's a twist: **nothing we do will have an immediate result**. The slow pace at which the planet's temperature adjusts to perturbations, or Earth's *thermal inertia*, means that we're only now seeing the temperature results from environmental changes from twenty or thirty years ago. Even if we were to stop putting out *any* human-caused greenhouse gases today, thermal inertia would guarantee that we'd still see temperature increases for at least the next twenty years. Add to that the feedback effects from environmental changes that have already happened: ice cap losses increasing polar ocean temperatures, accelerating overall warming; melting permafrost in Siberia releasing methane, which can be up to 72 times more powerful a greenhouse gas than carbon dioxide; overloaded carbon sinks in oceans and soil losing their ability to absorb CO2. These factors combine in a way that could make even our best efforts too slow to avoid disaster.

This means that we're looking at an increase of at least another half degree or so, guaranteed. If plans like stabilization wedges take another decade to get going, that's roughly another quarter-degree increase on top if it, *at minimum*. And if the emissions reduction plans don't work as well as intended or see technical or political delays, or if we have faster-than-expected temperature increases, we may end up hitting that one degree increase despite taking all the right steps now.

Now, it's possible that one degree is insufficient to push us into a climate tipping point. It's also possible that a brief period at "tipping point" temperatures may not be enough to make the change stick — so that we hit the danger point, but efforts long underway pay off at the right time to pull us back from the brink.

But if we *do* face an incipient global climate disaster, we have one last card to play: geoengineering.

PUTTING CLIMATE CHANGE ON HOLD

Geoengineering is the science fiction-sounding idea that we can make large-scale adjustments to the geophysics of our home planet in order to hold off environmental disaster: blocking a fraction of incoming sunlight with giant mirrors in space or fine particles in the stratosphere; using genetically-engineered plants and microbes to remove atmospheric carbon dioxide and methane at a faster-than-natural rate; pumping seawater high into the atmosphere to increase the reflectivity of clouds; dumping iron into the oceans to stimulate the growth of carbon dioxide-devouring plankton; capturing CO_2 directly from the atmosphere, and more.

At first blush, you might think that this sounds like a way to solve global warming without having to change how we live — unfortunately, it doesn't work that way. At best, geoengineering would be a way to delay disaster, giving us a bit more time to implement the kinds of worldwide changes to our economies, our energy infrastructure, and our ways of life necessary to make it to the next century with our civilization intact. If we imagine global warming to be akin to a lifestyle-related disease — Type 2 Diabetes, for example — geoengineering would be the medicine used to suppress the worst symptoms, so that changes to diet and exercise have time to take effect.

And like many medicines, geoengineering wouldn't be without potentially negative side-effects. Few if any of the various geoengineering proposals would without the potential for consequences that could make an already-bad situation even worse. We still know too little about the nuances of geophysical systems to be confident in our ability to engineer changes without risking disaster. This alone is enough to make some otherwise technology-friendly environmentalists reject the notion of geoengineering out of hand.

Compounding the dilemma is the recognition that, as a process that would affect every nation and every person on the planet, political struggles and ethical debates will be part-and-parcel of any implementation of geoengineering. At best, geoengineering would be a trigger for loud debates on the floor of the United Nations and between television pundits; at worst, it would be a catalyst for war.

But we may be running out of alternatives.

Although we know broadly what we need to do to avoid climate disaster, the details of how we make the necessary changes happen — politically, economically, technologically, and ethically — are enormously complex. We can solve the problems of changing the world to avoid environmental catastrophe, if we have enough time. But that's a *big* if.

Failure has a price. As the climate deteriorated, we'd find ourselves forced to look for ways to mitigate the very worst impacts of global warming. No matter how swiftly we act, we'd still see disastrous events, and many deaths; in time, however, we'd learn how to deal with the new climate. Hopefully, we'd be able to do so before too many people died from heat waves, drought, opportunistic diseases, storms, resource wars, forced migration, and the like. But make no mistake: even a "good" mitigation scenario would still be catastrophic for many around the world.

That's why the geoengineering option appeals to many: systems that cool the planet a bit over the short run could suppress many of the more disastrous effects of warming temperatures, even as we continue with emissions reductions.

Prudence dictates that we need to act to transform how we live, but also prepare for what happens if our best efforts fail. Geoengineering solutions are drastic remedies for disastrous situations. And if we're forced to rely on geoengineering to give us time to change the world, we need to know what to do — and, perhaps more importantly, what *not* to do.

The early examination of options we'd rather not use is vital. *If* the climate collapses faster than expected, or *if* our efforts fail (or are blocked by recalcitrant leaders), we *will* see people desperate for survival trying out these kinds of last-ditch solutions. It would be a good idea if they knew what the consequences could be before they choose which one(s) to try. If the planet faces disaster, not everyone will be willing to simply throw up their hands and say, "well, we're boned!" — some of us will do what we can to catch human civilization before it's gone completely over the cliff. We'll try everything we can think of to forestall climate disaster.

Geoengineering Awareness

When I first started writing about geoengineering in 2005, it was, at best, a fringe idea, a concept out of the fever dreams of science fiction writers and former nuclear weapons specialists looking for a new way to control nature. Most folks who had heard of geoengineering dismissed it as crazy. Terms like "arrogance," "hubris," and "insanity" got tossed around by environmentalists and even many scientists. Opponents of the idea quickly focused on two big concerns:

1) The cranks and industry-funded pundits in loud opposition to doing anything about global warming would latch onto geoengineering as a way of dealing with the problem without changing how we live.

2) Our knowledge of the Earth's complex, interdependent systems is woefully incomplete, and any attempt to make big, system-level changes was almost certainly doomed to failure.

But over the past few years, the concept of geoengineering has started to gain legitimacy. This has come, in part, from highly qualified geoscientists looking closely at what geoengineering would take to carry off successfully. It has also come, in part, from the growing recognition that we're running out of time faster than we thought.

And it's turned out that, although there have been occasional non-scientific speculations that geoengineering would mean we wouldn't have to do anything else to deal with global warming,

---

* *Original version of this piece published at Open the Future in July, 2008*

the vast majority of people studying the subject have embraced the need for rapid, aggressive reductions of carbon emissions — and see geoengineering as the best hope for having enough time to make those reductions happen.

The question of complexity remains salient, however. We don't know enough about complex interactions of geophysical systems to make good predictions of how effective (or how safe) different geoengineering proposals would be. Unanticipated consequences are almost guaranteed with something as massive in scale as geoengineering. The need to monitor and be able to scale back or shut down a geoengineering project would be absolute.

There are those would dismiss any thought of geoengineering on the basis that it's wrong to conduct a planet-wide experiment in climate engineering. Unfortunately, this neglects the fact that we're *already* conducting a planetary climate experiment, only we've lost the lab notes, don't have a control, and got massively drunk the night before. We've dumped massive amounts of garbage into the atmosphere with little consideration of the long-term results. Now we get to see what happens.

It's been fascinating to watch the evolution of the mainstream media coverage of the geoengineering concept. I'm actually pretty pleasantly surprised: most of the articles I've seen have had an overall tone of caution about the proposals, even while recognizing that if we end up using geoengineering technologies, it's because things have gotten so bad that we're down to our last-ditch methods of avoiding disaster. The basics of the stories have been pretty consistent: we're in an even bigger climate mess than we thought, so real scientists have begun to consider options for climate modification that they might have dismissed in the past, simply to head off catastrophe; nonetheless, more research needs to be done. I haven't seen any news stories (as opposed to opinion pieces, or blog articles) that even imply that geoengineering would be considered a replacement for de-carbonization, and I'm seeing fewer news articles that start from the perspective that such climate modification is inherently wrong, period.

Greg Lamb's 2007 article in the *Christian Science Monitor*, "**Can We Engineer a Cooler Earth?**" is an example of the changes to how the mainstream press covers of the concept. Yes, he quotes me in the piece, but he does so without too badly distorting what I tried to say. It does make me sound like more of an enthusiast than I really am, but it gets the essential point across: we need to get our carbon emissions down, but the climate is changing faster and harder than even the most pessimistic models had predicted a few years ago.

> *Blocking sunlight, adds futurist Cascio, "is at best a delay of the worst temperature-related consequences of global warming in order to give us more time for de-carbonization." [...]*

> *Schemes to slightly dim sunlight also wouldn't solve the problem of ocean acidification, caused by airborne CO_2 entering seawater. More-acidic oceans would harm coral reefs and upset ocean ecology, with possible far-reaching effects. Ocean acidification is "at least as big" a problem as that of CO_2 in the air, Cascio says.*

That last point is critical. Even if we manage to avoid a heat-related crisis with geoengineering, we'll still need to eliminate our industrial carbon emissions as quickly as possible to avoid ocean ecosystem collapse. I suspect that we'll come to see ocean acidification as a *bigger* problem than atmospheric warming, in fact.

If a geoengineering attempt is (as I suspect) highly likely in the next decade or two, we damn well should know a bit more about what we're doing. We need to have a major research project already underway to figure out which of our options are likely to have the best results at the least-disastrous costs. We need to be able to warn people off of the really terrible options by being able to point them to the less-bad (and potentially helpful) alternatives. Fortunately, this will require a great deal more knowledge about geophysical systems, knowledge that will prove beneficial even if we manage to avoid the more desperate solutions.

To paraphrase Stewart Brand, we are as planetary engineers, so we may as well get good at it.

EXAMPLES OF GEOENGINEERING TECHNIQUES

Orbiting Space Shield

Cloud Brightening
with Sea Water

Stratospheric Sulfate Injection

Genetically Modified
Trees

Ocean Fertilization
Triggering Algae
Blooms

Carbon Capture
and Burial

BLOCKING/REFLECTING SUNLIGHT TO REDUCE TEMPERATURES:

Orbiting Space Shield
Stratospheric Sulfate Injection
Cloud Brightening

REMOVING CARBON FROM THE ATMOSPHERE:

Genetically Modified Trees
Carbon Capture and Burial
Ocean Fertilization

COMPARING GEOENGINEERING TECHNIQUES

In the 2009 paper **"The Radiative Forcing Potential of Different Climate Geoengineering Options,"** published as a draft in the open access scientific journal *Atmospheric Chemistry and Physics*, Tim Lenton and his student Naomi Vaughn of the Tyndall Centre for Climate Change Research and the University of East Anglia, UK give us one of the first useful comparisons of different geoengineering techniques.

Lenton and Vaughn focus strictly on the radiative impact of geoengineering — that is, how much heat absorption is prevented — and don't examine costs or risks. Their goal is to help figure out the "benefit" half of the cost-benefit ratio. (Lenton and Vaughn have another paper in the works taking a look at the cost side, and that is likely to be just as important as this one.)

Lenton and Vaughn split geoengineering proposals into two categories:

Shortwave options that either increase the reflectivity (or *albedo*) of the Earth or block some percentage of incoming sunlight. These include mega-scale projects like orbiting mirrors and stratospheric sulfate, as well as more localized and prosaic methods like white rooftops and planting brighter (=more reflective) plants.

Longwave options that attempt to pull CO_2 out of the atmosphere in order to slow warming. These include massive reforestation projects, "bio-char" production and storage, various

---

* *Original version of this piece published at Open the Future in January 2009*

air capture and filtering plans, and ocean biosphere manipulation with iron fertilization or phosphorus.

Lenton and Vaughn run the numbers on the likely maximum results from each of the methods, working under the assumption of simultaneous aggressive carbon cutting efforts. Their numbers are a first-order approximation, but sufficient for useful comparison. One thing that becomes immediately clear is that **no form of geoengineering would be enough to avert catastrophe if emissions aren't cut quickly**. Unfortunately, they also argue that **even aggressive carbon emission cuts won't be enough to forestall disaster alone, either**.

So, what works?

Technique	Overall Effect	Other Issues
Space Shield	Very High	Very expensive, outside present technological ability
Stratospheric Sulfate Injection	Very High	If stopped abruptly, temperatures spike
Cloud-Brightening	Very High	Uncertainties about how to make it happen
Increased Desert Albedo	High	Environmentally devastates deserts
Air Capture and Storage	High	Uncertainties about how to make it happen
Reforestation	Moderate	Likely reduction in land available for crops
Bio-Char	Moderate	Low at first, increasing over (long) time
Ocean Fertilization	Low	Increases over time, question of ocean environment impact

Most effective (again, strictly in terms of radiative impact) over this century would be some kind of temperature moderation through space shields, stratospheric injection, or increasing cloud levels with seawater. Any of these, alone, could actually be enough to counteract global warming **along with aggressive carbon emission reductions**.

Next would be increasing desert albedo (essentially putting massive reflective sheets across the deserts of the world) or direct carbon capture and storage (ideally captured from burning biofuels). These would slow global warming disaster, but wouldn't necessarily be enough to stop it. Bio-char, reforestation, and increasing cropland & grassland albedo come in third, half again as effective as the previous proposals; the remaining methods would be even less effective, in some cases multiple orders of magnitude less-effective.

And all of these proposals have drawbacks. Space shields would be ridiculously expensive for the foreseeable future. Stratospheric injection alters rainfall patterns, and any abrupt cessation of albedo manipulation would be worse than what had been prevented. Laying thousands upon thousands of square kilometers of reflective sheets across the desert is an ecosystem nightmare, while reforestation at sufficient levels to have an impact — and any kind of biofuel or cropland/grassland modification — would be incompatible with feeding the Earth's people. Carbon capture has the fewest potential drawbacks, other than cost — and the fact that it alone wouldn't be sufficient to stop disaster, only delay it.

With the various drawbacks (which Lenton and Vaughn will examine in more detail later this year), why even consider geoengineering?

The explanation comes as an extension of the "bathtub model" discussed by Andy Revkin in his Dot Earth column in the *New York Times*.

Imagine the climate as a bathtub with both a running faucet and an open drain. As long as the amount of water coming from the faucet matches (on average) the capacity of the drain, the water level in the tub remains stable. This is a metaphor for the carbon cycle — as long as the carbon production (from biology, from geology, and even from industry) more-or-less matches the capacity of carbon sinks, we're okay.

Over the course of the last couple of centuries, however, we've been turning up the water flow — increasing atmospheric carbon concentrations — at first slowly, then more rapidly. At the same time, the drain itself is starting to get clogged — that is, the various environmental carbon sinks and natural carbon cycle mechanisms are starting to fail. With a bathtub, the combination of a clogged drain and a wide-open faucet inevitably results in water spilling over the sides of the tub. With the global carbon cycle, the combination means a catastrophic shift in climate.

The value of this model is that it illustrates nicely some of the less-intuitive aspects of the climate dilemma. We can see, for example, that simply slowing emissions to where they were (say) a couple of decades ago won't necessarily be enough to stop a disaster, if the carbon input is still faster than the carbon sinks can handle.

So far, our efforts at stopping this catastrophe have — rightly — focused on reducing the water flowing from the faucet (cutting carbon emissions) as much as possible. But the flow of the water is still filling the tub faster than we can turn the faucet knob (we're far from getting carbon emissions to below carbon sink capacity). Without something big happening, we're still going to see a disaster.

The *shortwave* geoengineering proposals, by blocking some of the incoming heat from the Sun, are the equivalent here to building up the sides of the tub with plastic sheets. The tub will be able to hold more water, although if the sheeting fails, the resulting spillover will be even worse than what would have happened absent geoengineering.

The *longwave* geoengineering proposals, by increasing carbon capture, are the equivalent here to clearing out the drain, or even drilling a few holes in the bottom of the tub (let's assume that just goes to the drain, too). The water will leave the tub faster, but you may have to drill a lot of holes to have the impact you need — and drilling too many holes could itself be ruinous.

According to Lenton and Vaughn's study, the longwave geoengineering proposals would be much more effective in the long run — at millennium scales — than shortwave, but the shortwave would have a more immediate impact. It's likely that a combination of the two approaches would be most effective overall, of course coupled with aggressive carbon emission reductions. Build up the sides of the tub *and* drill a few holes, in other words, even as you frantically try to close the tap.

Of all of the longwave proposals, air capture seems to be closest to a winner here, but the costs (and technology) remains a bit unclear, and will take some time to get up and running in any event. That delay will mean added pressure to use one or more of the shortwave approaches. My guess is that stratospheric sulfate injection will be cheaper at the outset than the cloud albedo manipulation with seawater, but the latter seems likely to have fewer potential risks; we'll likely try both, but probably transition solely to cloud manipulation. The various minor proposals — reforestation, urban rooftop albedo, and the like — certainly won't hurt to do, and every little bit helps, but alone are likely insufficient.

Lenton and Vaughn's study is precisely the kind of research that is needed to better understand what the geoengineering options are. As I emphasize at every turn, this doesn't obviate the need for aggressive reductions in carbon emissions. But it's looking more and more like simply changing our light bulbs, boosting building efficiency, and taking a bike instead of a car, while clearly helpful, will still be insufficient to avert disaster, and even a global shift away from fossil fuels wouldn't come in time to stop the water from spilling over the edges of the tub.

Nature stopped being natural centuries ago. It's been in our hands, under our influence, for much longer than we've been willing to admit. We've got to get smart about how we're reshaping the environment — and do so before it's too late.

THE QUESTION OF METHANE

One climate wild-card may force us to engage in geoengineering far sooner — and using more radical techniques — than we might otherwise want.

In the Siberian arctic, the permafrost appears to be melting. This is happening due to a combination of natural Arctic temperature cycles, global warming (Siberia is warming faster than any other place on Earth), and a feedback effect from melting snow — the darker ground absorbs more heat, resulting in faster melting of adjacent permafrost. Siberian permafrost covers a million square kilometers of ground that's largely peat bog; the peat has been producing methane for centuries, but that methane has been trapped under the permafrost. With the permafrost melting, the methane would be released into the atmosphere, accelerating global warming by a substantial amount. How quickly the methane would be released remains an open question — would it take years to release it all? Decades? A century or more?

There are still many questions about the speed and scale of the permafrost melt, and the amount of methane being released. But the preliminary studies seem to agree that a methane release is underway, and could get bigger quickly.

There are two important things to know about this situation: the amount of methane that would be released is projected to be in the multi-gigaton range — one source says 70 billion tons, another

---

* *Original version of this piece published at Worldchanging.com in August, 2005.*

says "several hundred" billion tons; and methane is at minimum 20-25 times more powerful a greenhouse gas than carbon dioxide over the course of a century. But methane cycles through the atmosphere in just a few decades, meaning that actual intensity of methane's impact is closer to *70* times CO_2 during its stay in the air. In essence, the release of (say) 100 billion tons of methane would be the heat-trapping equivalent of up to **7 trillion tons of CO_2**. To put that number into perspective, the total annual output of greenhouse gases from the US is about 7 **billion** tons of CO_2 equivalent.

This is a big deal.

But there's actually a third important thing to know: although CO_2 takes upwards of a century to cycle out of the atmosphere naturally, methane (CH_4) takes only about ten years. Why the difference? Chemical processes in the atmosphere break down CH_4 (in combination with oxygen) into CO_2+H_2O — carbon dioxide and water. In addition, certain bacteria — known as *methanotrophs* — actually consume methane, with the same chemical results. These processes have their limits, however; an abundance of methane in the atmosphere can overwhelm the oxidation chemistry, making the methane stick around for longer than the typical 8-10 years, and the commonplace methanotrophic bacteria evolved in an environment where methane emerges gradually.

These are pretty much the only two natural methane "sinks." There are a few small-scale human processes that can make use of methane (for the production of methanol for fuel, for example) and function as artificial sinks, but such efforts would be hard-pressed to capture methane released across nearly a million square kilometers. This, then, is where we start to consider the option of planetary engineering.

Both of the natural processes are, in principle, amenable to human intervention. The oxidation of methane into CO_2 and water is a well-understood phenomenon, and relies on the presence of OH (hydroxyl radical); upwards of 90% of lower atmosphere methane is oxidized through this process. But OH is something of a problem chemical, in that it's also a key oxidation agent for many atmospheric pollutants, such as carbon monoxide and NO_x. Although we could produce OH to enhance the natural chemical oxidation process, the side-effects of pumping enough OH into the atmosphere to oxidize all of that methane would be unpredictable, but almost certainly quite bad.

So what about methanotrophic bacteria? Such bacteria have long been recognized in freshwater areas and soil, and have had limited use in bioremediation efforts. Methanotrophic *Archaea* — similar to bacteria, but a wholly different kingdom of organism — were recently identified in the oceans; research suggests that methanotrophic *Archaea* may be responsible for the oxidation of up to 80% of the methane in the oceans. Methanotrophic microbes can also be temperature extremophiles, as they were among the CO_2 after the Larsen B ice shelf collapsed.

We recently began to learn much more about how methanotrophic bacteria function, as a team from the Institute for Genomic Research sequenced the genome of the methanotroph

Methylococcus capsulatus. The scientists discovered that *Methylococcus* has the genomic capacity to adapt to a far wider set of environments than it is currently found in. They also looked at the possibility of enhancing the microbe's ability to oxidize methane, although admittedly for purposes other than straight methane consumption.

Freshwater methanotrophs are increasingly well-understood, but present a limited means of methane remediation. Methanotrophic *Archaea* have demonstrated ability to act as a major methane sink, at least in the oceans, and to live in extreme temperature conditions. Neither is a good fit for Siberia. The Siberian arctic, while warming, remains damn cold, but the melted permafrost lakes will be freshwater settings.

It appears to me that what will be the most effective means of mitigating and remediating the gargantuan methane excursion from the Siberian permafrost melt would be using genetically-modified forms of methanotrophic bacteria, with greater oxidation capacity and the *Archaea*-derived resistance to extreme cold (these may well go hand-in-hand, as one way that deep sea methanotrophs survive the icy depths is through internal energy production from methane consumption). Given the size of the region, we'll need lots of them, but that's another advantage of biology over straight chemistry: the methanotrophs would be reproducing themselves.

It's unlikely that abundant reproduction of genetically-modified methanotrophs would pose a larger risk — at the very least, they'd be limited to the post-permafrost lakes, as they'd be based on freshwater-only species — and a mass of methanotrophic organisms would undoubtedly be helpful for reducing overall atmospheric methane beyond the Siberian release. More importantly, the successful introduction of such organisms would give us practice for what would be a far, far greater problem: the undersea methane clathrates, which are believed to contain upwards of 500 billion tons of CH_4. Undersea clathrate melts have been implicated in mass extinctions in the geologic past; the significant climate warming that would result from an unmitigated Siberian release would pale in comparison to the effects of a clathrate melt.

What are the outstanding questions we need to answer before we could consider creating GMO methanotrophs?

Is it physically possible? Could a sufficient number of methane-eating bacteria even be produced to counter a fast release of methane from the Siberian bogs?

Is it biologically possible? Could a species of methanotrophic bacteria be engineered to be able to thrive in Siberian conditions and consume large quantities of methane?

What are the unrecognized risks? What are we missing in an initial risk analysis? Saying "we don't know the risks" doesn't, in and of itself, mean "we should not attempt this," it means "we need to do more research." Clearly, if the risks from enhancing the methane consumption and environmental adaptation capacities of a methanotroph could lead (through species-hopping genes or simple mutation) to even harder-to-manage problems than gigatons of atmospheric methane, this isn't an option. Boosting

OH levels in the region would be the fallback position, as we have more experience with managing CO and NO_x pollutants.

We may be forced to do something like this. With Siberian methane, the more cautious options are extremely limited. We're no longer in a position to stop the melting, even by ceasing all greenhouse gas production today; the temperature increases we're seeing now are the results of greenhouse gases put into the atmosphere decades ago.

In a way, among the different scenarios forcing us to consider "terraforming," this is probably close to the best choice. Failure would be drastic, but not utterly catastrophic (unless the resulting warming, in turn, melts the undersea clathrates, at which point all bets are off). And, as noted, this would allow for better refinement of technique and understanding of choices in the face of a similar-but-greater in magnitude problem down the road (in this case, the aforementioned clathrates).

A further advantage is that this is a process that could begin after we start to see significant methane output and could still have a measurably positive result. Using microbes for bio-"scrubbing" of methane from the atmosphere would work on methane that was a decade old as readily as methane fresh from the bog. We'd still see some effect from the methane that makes it to the atmosphere, but eventual removal would help to reduce that effect. This means that, should we face a situation where questions still need to be answered before we could comfortably begin to use the GMO methanotroph option, but we're starting to see an impact from the Siberian release, we don't necessarily have to rush past our better judgment in response. With a process of this magnitude, it's worth taking the time to get it right.

If we see a massive methane release, we would be facing a problem of a scale with few precedents in human history. No society on the planet would be unaffected; if left unmitigated, it would continue to affect the lives of our children, and our children's children, and generations beyond that. Our choices are few, and the risk of not acting is (potentially) immense. We may well be on the brink of a new era in planetary management. Let's hope we're up to the challenge.

GEOENGINEERING AND POLITICS

Global Climate and Global Power

I was pinged recently by the UK outfit Forum for the Future, a foresight team specializing in sustainable futures. They wanted to know what I thought would be the key issues the world would be confronting in 2030. "Climate" is the first thing that popped to mind, unsurprisingly, and we talked for a bit about what that might look like.

One possibility that stands out is the potential for a political shake-up coming from how we respond to climate disruption. We must not underestimate just how disruptive it's going to be to transform our lives and societies, both locally and globally. Over the next twenty years, it's highly likely that we'll have seen some significant changes in how we govern the planet.

This scenario most likely to make this apparent is one in which we embark upon a set of geoengineering-based responses to the climate problem, probably starting in the early-mid 2010s, as a way of supporting the carbon emission-reduction measures. These would likely be various forms of thermal management, such as stratospheric sulfate injections or high-altitude seawater sprays, but might also include some form of carbon capture, or even something not yet fully described. I see the mid-2010s as a probable starting period mostly out of a combination of global desperation and political compromise; geoengineering advocates might see it as already too late, while opponents would likely want to have more time to study models.

---

* *Original version of this piece published at Open the Future in December, 2008.*

As a result, by 2030, while various carbon mitigation and emission reduction schemes continue to expand, a good portion of international diplomacy would concern just how to control (and deal with the unintended consequences of) climate engineering technologies. It's not impossible that there will be an outbreak or two of violence over geoengineering management. I wouldn't be surprised if at one point, the world ceases geoengineering, only to find temperatures bouncing back up quickly; the process would then almost certainly be resumed.

This is a challenging world, and not just because of conflicts over control or the potential for unexpected impacts. It's a world in which the two familiar models of power — "hard" military power and "soft" cultural power — don't adequately describe the arena of competition. Although geoengineering might have the potential to be used harmfully, it would be insufficiently visible, swift, and controllable to serve as a broadly useful form of force; similarly, the memetic elements of a geoengineering strategy are keenly focused on scientific debates over uncertain results, a form of discourse which tends to be opaque to most citizens.

And the struggles over geoengineering wouldn't be happening in a vacuum. Over the next couple of decades, we'll be dealing with multiple complex global system breakdowns, from the present financial system crisis to peak oil production to the very real possibility of food system collapse. Climate disruption, with or without geoengineering, clearly falls into the category, as well: systems in which neither hard nor soft power work very well. All of these problems demand greater information analysis, long-term thinking, and accountability than traditional forms of power tend to offer.

The era of overlapping system threats is now clearly underway, and geoengineering will be a highlight of that period. New patterns of international behavior will almost certainly have emerged by 2030. My gut sense is that they'll have a strong legalistic component; in particular, one of the major points of debate over geoengineering will be liability for negative consequences. Given the need to deal with these overlapping crises, we might imagine the third form of power (beyond hard and soft power) as a kind of "administrative" power. (There's an intentional echo here of Thomas Barnett's "sysadmin force" concept, but this isn't meant as a direct link.) Although much of what I've been discussing here about administrative power focuses on the actions of states and transnational entities, the same concept could easily be applied to bottom-up groups and movements (just as hard and soft power concepts operate at both ends of the scale).

Administrative power is an admittedly boring name, and doesn't fit with the hard/soft dichotomy; perhaps "flexible power" works better. But no matter what you call it, over the course of the next few decades, we're likely to see the rise of an alternative model of competition that works directly with complex interconnected global systems. Geoengineering won't be the cause of it — really, the emergence of administrative power (or whatever you call it) is already underway — but could well be the action that makes this model of power clearly visible.

THE POLITICS OF GEOENGINEERING

Given the increasing visibility of geoengineering proposals, debates among scientists, environmentalists, and engineers are not hard to find. But these debates center on the scientific risks and merits of the re-terraforming proposals. Few people, regardless of position, have focused on a fundamental *non*-geophysical risk of the method: political control, costs, and stability.

To put it bluntly, global-scale efforts don't happen without global-scale reactions. Should we see geoengineering efforts, there will certainly be struggles over control of the program(s), conflicts over liability for problems, and — most troublingly — independent. "rogue" geoengineering projects undertaken in defiance of established guidelines.

CONTROL

Of the three kinds of political dilemmas regarding geoengineering, this is probably the easiest to grasp.

The question of control over geoengineering parallels to a surprising degree the question of control over (or legitimacy of) warfare: both emerge from considerations of a nation's ability to survive. It's reasonable to assume that the United Nations would expect to authorize and provide oversight for any re-terraforming project: the benefits would be transnational, so the costs should arguably be spread; the risks are transnational as well, so international oversight helps to defray blame; and given the scale

---

* *Original version of this piece published at Open the Future in October 2007.*

of such projects, nations that would be affected one way or another would demand consultation.

But like warfare, it's entirely possible that a state with the capacity to undertake such a project independently might decide that international restrictions are irrational, or that its survival is so threatened that the bureaucracy of a transnational body is unacceptable. Smaller nations following such a course would be declared "rogue nations" (and are addressed below); when a hegemonic nation does it, such as the United States or China, there may be little the international community can do in response.

Little, unless rival hegemonic powers come to believe that such independent geoengineering efforts threaten their security and environmental survivability. Then, like any other security threat, this could be a trigger for war.

LIABILITY

There's very little doubt that any geoengineering efforts begun without sufficient study could have a significant chance of triggering unforeseen results, simply due to the complexity of the geophysical systems involved. In a situation of imminent risk of (say) Greenland's ice sheet collapsing into the ocean, such unforeseen results may be an acceptable trade-off for avoiding certain disaster (to be clear, I don't think we're likely to see Greenland's ice sheet showing signs of imminent collapse within the decade, but it's precisely the kind of situation that would push even opponents of geoengineering to consider its use). But geoengineering strategies can have dangerous externalities. Take the solution geoengineering specialist Dr. Ken Caldeira suggests for discussion in an editorial in the *New York Times*:

What can be done? One idea is to counteract warming by tossing small particles into the stratosphere (above where jets fly). This strategy may sound far-fetched, but it has the potential to cool the earth within months. [...] If we could pour a five-gallon bucket's worth of sulfate particles per second into the stratosphere, it might be enough to keep the earth from warming for 50 years. Tossing twice as much up there could protect us into the next century.

Sulfate particles can be found in the atmosphere as the result of volcanos and human industrial activity, and can measurably reduce radiative forcing. Sulfate particles are not benign, however, and can be linked to a variety of human diseases, as well as to acid rain and changes to cloud formation. Sulfate injection would put the particles in the stratosphere, ostensibly avoiding such problems — assuming all goes as planned. Unfortunately, given a program of the magnitude of stratospheric sulfate injection, mistakes, accidents, and even malice can't be ruled out.

Such problems resulting from sulfate particles would not be limited to the nation or nations leading the project; neither is it likely that they'd be distributed evenly around the globe. It's highly likely, instead, that the problems resulting from pumping five gallons of sulfate particles *per second* into the atmosphere for an extended period would affect some nations more than others. If history is any guide, the nations most likely to bear the burden of these health and environmental problems are those that are the poorest and least stable. Will the countries and institutions pushing for the geoengineering strategy be equally as eager to pay for reparations and recovery?

This assumes that there are no wildly destructive unanticipated side-effects, in which case the financial, environmental and human costs could be dramatically higher.

(This raises the interesting possibility that the insurance industry — especially the re-insurance companies — may exert tremendous pressure on governments and institutions not to adopt a geoengineering strategy as anything but a final fall-back.)

Rogue Projects

The first two political dilemmas arising from geoengineering efforts are heightened versions of relatively conventional international issues: political control and distribution of costs. The third dilemma, conversely, has few precedents.

It is possible that, should the international community refrain from geoengineering strategies, one or more smaller, non-hegemonic, actors could undertake geoengineering projects of their own. This could be out of a legitimate fear that prevention and mitigation strategies would be insufficient, out of a disagreement with the consensus over geoengineering safety or results, or — most troublingly — out of a desire to use geoengineering tools to achieve a relative increase in competitive power over adversaries.

It's entirely possible, even likely, that the hegemonic international powers will decide, after careful study, that the potential risks of substantive geoengineering outweigh the potential benefits, and that no such strategies should be pursued. However, we know that the negative impacts of global warming are distributed unevenly, and what may be acceptable levels of climate disruption for the major states may be utterly devastating for poorer, smaller nations. It is in this context that a scientifically-powerful developing nation — India or Brazil, for example — may decide that it is unwilling to abide by UN decisions about re-terraforming, and begin to undertake such a strategy.

It may have concluded that the impacts of climate change would hit it too quickly for carbon reductions around the world to have an effect; it may see geoengineering as its only choice. Conversely, it may have concluded that the scientific arguments against geoengineering were faulty, and that such an effort could be undertaken safely, regardless of the success of other solutions. Would this rogue effort be backed up by a threat to use all means necessary to defend the project? Would the UN or the hegemonic powers be willing to use sanctions, interdiction of project-related materials, even war to stop the rogue?

Moreover, with the geoengineering technologies on the table, there's no guarantee that they'll only be used for environmental purposes. Nuclear proliferation and open-source warfare (as seen in Iraq and Afghanistan) have made successful conventional warfare far more difficult — perhaps even functionally impossible. Geoengineering weapons may offer a new potential to disrupt one's enemies over a long, subtle campaign. Is this likely? Probably not, but it's sufficiently possible that governments will be forced to consider it.

No nation that sees itself as a great power is going to be willing to risk having its climate and environment completely in the hands of another nation. Research into methodologies for geoengineering will happen simply out of self-preservation — after all, nobody wants to fall victim to a "terraforming gap."

IMAGINING THE UNTHINKABLE

Finally, we have to recognize that the "rogue actors" *need not be states*. While the costs of geoengineering strategies may be enormous, they wouldn't necessarily be out of the range of some of the global billionaires. The movie scenario's not hard to imagine:

In a world on the verge of destruction... while nations delay and scientists bicker... one man sees a way to save us all.

"But the UN isn't sure it's safe!"

"To hell with the UN! I just want to save the Earth!"

As the planet burns, Warren Gates-Branson III crosses the line no nation dares cross.

"All my money counts for nothing if the world's gone to hell!"

He is... THE TERRAFORMER.

Unfortunately, the less-heroic version's not hard to imagine, either: a cadre of scientists and engineers willing to say anything to test out pet ideas, a multi-jillionaire who believes himself smarter than those bureaucrats at the UN, and a planet already on the tipping point of catastrophe, just waiting for some kind of event to trigger an unstoppable cascade of environmental tragedy.

When I argue that we need to start studying geoengineering *now*, I don't simply mean the climate scientists and geophysicists. I mean everyone who worries about policy, embraces activism, works with NGOs and movements, or considers herself or himself a stakeholder in the well-being of the planet. If we ignore this possibility, decisions will be made without our consent, even without our knowledge. We need to understand the kinds of choices we'll face if we continue to delay action on global warming. Geoengineering might, with the wrong moves, be catastrophic; it might, with the right knowledge and technologies, be our final hope. But it must not be a decision made by ideology, or as a military maneuver, or out of convenience.

Terraforming War

Preventing global warming from becoming a planetary catastrophe may take something even more drastic than renewable energy, superefficient urban design, and global carbon taxes. Such innovations remain critical, and yet disruptions to the Earth's climate could overwhelm these relatively slow, incremental changes in how we live. As reports of faster-than-expected climate changes mount, a growing number of experts worry that we might ultimately be forced to try something quite radical: geoengineering.

Geoengineering involves humans making intentional, large-scale modifications to the Earth's geophysical systems in order to change the environment. These can include sequestering atmospheric carbon dioxide in the oceans, changing the reflectivity of the Earth's surface, and pumping particles into the stratosphere to block a fraction of incoming sunlight. Many of these proposals mimic natural events, so we know that — in principle — they can work, although there is insufficient understanding of their potential side effects. Unsurprisingly, geoengineering is highly controversial, and even proponents view it as a "Hail Mary" pass, to be considered only after all other options have failed.

But geoengineering presents more than just an environmental question. It also presents a geopolitical dilemma. With processes of this magnitude and degree of uncertainty, countries would

---

* *Originally published as "Battlefield Earth" at ForeignPolicy.com, January 27, 2008*

inevitably argue over control, costs, and liability for mistakes. More troubling, however, is the possibility that states may decide to use geoengineering efforts and technologies as weapons. Two factors make this a danger we dismiss at our peril: the unequal impact of climate changes, and the ability of small states and even nonstate actors to attempt geoengineering.

For a variety of political and natural reasons, global warming affects some countries differently than others. Fragile economies and weak infrastructures tend to worsen the results of climate disruptions, a problem exemplified by Bangladesh's vulnerability to monsoons, accelerating desertification in northern China, and, most visibly, Hurricane Katrina's devastation in New Orleans. At the same time, warming and altered rainfall patterns may — temporarily — improve conditions for countries in extreme latitudes, increasing harvests in Canada and Russia for a few years. Similarly, intentional changes meant to fight global warming would also have differential results.

At the same time, the resources required for geoengineering projects can vary dramatically. A start-up company called Climos and the government of India have each begun to prepare tests of "ocean iron fertilization" to boost oceanic phytoplankton blooms, in order to extract carbon dioxide from the atmosphere, at a cost of just a few million dollars. At the other end of the spectrum, projects like the injection of megatons of sulfur dioxide into the upper atmosphere to simulate the effects of a volcano would easily cost in the tens of billions of dollars — still within the means of most developed countries.

It's this combination of differential impact and relatively low cost that makes international disputes over geoengineering almost inevitable. Even if there is broad consensus that geoengineering is too risky, research into environmental modification will happen simply out of self-preservation — nobody wants to fall behind. Moreover, it's not hard to imagine some international actors seeing geoengineering as something other than solely a way of avoiding environmental disaster.

It wouldn't be the first time states looked at the environment as a weapon. In the early 1970s, the Pentagon's Project Popeye attempted to use cloud seeding to increase the strength of monsoons and bog down the Ho Chi Minh Trail. In 1996, a group of Air Force and Army officers working with the Air Force 2025 program produced a document titled "Weather as a Force Multiplier: Owning the Weather in 2025" (it never went anywhere). The Soviet Union reputedly had similar projects underway. But although the idea of a geoengineering arms race may superficially parallel this line of thinking, it's actually a very different concept. Unlike "weather warfare," geoengineering would be subtle and long term, more a strategic project than a tactical weapon; moreover, unlike weather control, we know it can work, since we've been unintentionally changing the climate for decades.

The offensive use of geoengineering could take a variety of forms. Overproductive algae blooms can actually sterilize large stretches of ocean over time, effectively destroying fisheries and local ecosystems. Sulfur dioxide carries health risks when it cycles out of the stratosphere. One proposal would pull cooler water from the deep oceans to the surface in an explicit attempt to shift the trajectories of hurricanes. Some actors might even deploy counter-geoengineering projects to slow or alter the effects of other efforts.

The subtle, long-term aspects of geoengineering could make it appealing. Because the overt purpose of geoengineering would be to fight global warming, and because complex climate systems would make it hard to definitively blame a given project for harmful outcomes elsewhere, offensive uses would likely be hard to detect with certainty. And, in a world where nuclear deterrence remains strong but the value of conventional military force has come under question, states will look for alternative, unexpected ways of boosting their strategic power relative to competitors.

Despite the global impact of geoengineering, the differential climate patterns and the resilience of local technological, economic, and social infrastructures guarantees that some states will fare better than others. Much as Cold War nuclear strategists could argue about "winning" a nuclear war by having more survivors, advocates of a Global Warming War might see the United States, Western Europe, or Russia as better able to "ride out" climate disruption and manipulation than, say, China or the countries of the Middle East. It's a new version of "thinking the unthinkable."

Smart policies could lessen these risks. The 1977 Environmental Modification Convention, produced by the United Nations in response to Project Popeye, prohibits the use of engineered weather or environmental changes for military purposes; signatory countries may wish to look at ways of monitoring and enforcing this treaty. Outright banning of geoengineering research is highly unlikely, as it offers a last-ditch hope for staving off climate disaster. Instead, putting research into the hands of transparent, international bodies could reduce the temptation to "weaponize" geoengineering; internationalization could also help to spread the liability and costs, reducing one potential source of tension.

The best strategy to avoid the possible offensive use of geoengineering techniques, however, is twofold: First, embrace the social, economic, and technological changes necessary to avoid climate disaster before it's too late; and second, expand the global environmental sensor and satellite networks allowing us to monitor ecosystem changes — and manipulation. This strategy may not reduce the temptation to look at geoengineering as an offensive capacity, but it would ensure that experiments and prototype efforts couldn't readily be hidden under the cover of fighting climate change. We know all too well that the international contest for power will continue even in the face of a growing global threat. It would be a tragedy if, in seeking to avoid environmental catastrophe, we inadvertently enabled a new quest for geopolitical advantage. The risks of turning the Earth itself into a weapon are far too great.

GEOETHICS

GEOETHICAL PRINCIPLES

The pace and course of global warming-induced climate disruption is such that, even with an aggressive global effort to cut greenhouse gas output starting today, temperatures will continue to rise for two or three decades. If the effect of rising temperatures hits a "tipping point" resulting in far-more-radical changes to the Earth's ecosystems than one might otherwise expect, we may be forced into using riskier, planetary-scale engineering projects to mitigate the changes and return us to "Earth-like" conditions.

But whether we end up taking the mitigation or the adjustment course, we will want — *need* — clear guidelines to help us make the right choices. Such guidelines would, for some, seem like common sense; indeed, their use would not be to tell us what to do, but as a consistent metric against which to test proposals. These principles would not tell us whether a given strategy would succeed or fail, but whether the strategy would be the right course of action.

As an explicit parallel to bioethics, these guidelines would be known as "geoethics."

Bioethics are the guidelines against which biomedical researchers and practitioners measure their own difficult decisions. While the concept is by no means new, it was first formalized in 1979, in a book entitled *Principles of Biomedical Ethics* by Tom Beuchamp and James Childress. Beuchamp and Childress conceived four core principles: *autonomy,* the personal

---

* *Original version published at Worldchanging.com in July, 2005.*

73

responsibility over our own lives, and the ability to make decisions for ourselves; *non-maleficence*, essentially "first of all, do no harm" (a notion derived from Hippocrates, but not actually part of the Hippocratic Oath); *beneficence*, a positive obligation to advance the welfare of others; and *justice*, the allocation of healthcare resources according to a just standard. These have become widely-accepted core principles for many working in the medical practice and medical research fields.

Like bioethics, the term "geoethics" is not new; unlike its biological cousin, there is no consistent definition of what geoethics covers, let alone its core principles. The closest I've found comes from Mike Treder at the Center for Responsible Nanotechnology, in an essay about the 1st Annual Workshop on Geoethical Nanotechnology. Treder defined geoethics thusly:

"Geoethical" means widely agreed-upon principles for guiding the application of technologies that can have a general environmental (including people) impact, much like bioethical principles (autonomy, beneficence, nonfeasance, justice) guide the application of curative technologies that specifically impact one or more patients.

Suggestive, but still vague. How is "technology" defined — would cars be included? Highways? Cities? Fire? What about practices that are not explicitly technological, but demonstrate an observable environmental impact (such as deforestation, agriculture, and mining)? How much of an environmental impact is enough to be covered? Subsequent literature searches only muddied the waters further.

The upside of this lack of consistency is that we can define geoethics and geoethical principles for ourselves without too much worry about disagreement with an established definition. Here's a draft definition:

Proposed phrasing:

Geoethics is the set of guidelines pertaining to human behaviors that can affect larger planetary geophysical systems, including atmospheric, oceanic, geological, and plant/animal ecosystems. These guidelines are most relevant when the behaviors can result in long-term, widespread and/or hard-to-reverse changes in planetary systems, although even transient, local and superficial alterations can be considered through the prism of geoethics. Geoethical principles do not forbid long-term, widespread and/or hard-to-reverse changes, but require a consideration of repercussions and so-called "second-order effects" (that is, the usually-unintended consequences arising from the interaction of the changed system and other connected systems).

Proposed core principles:

Interconnectedness — planetary systems do not exist in isolation, and changes made to one system will have implications for other systems.

Diversity — on balance, a diverse ecosystem is more resilient and flexible, better able to adapt to natural changes.

Foresight — consideration of effects of changes should embrace the planetary pace, not the human pace.

Integration — as human societies are part of the Earth's systems, changes made should take into consideration effects on human communities, and the needs of human communities should not be discounted or dismissed when considering overall impacts.

Expansion of Options — on balance, choices made should increase the number of options and opportunities for future generations, not reduce them.

Reversibility — changes made to planetary systems should be done in a way that allows for reconsideration if unintended and unexpected consequences arise.

Going into a bit more detail:

Interconnectedness is a recognition that the various planetary systems have deep and sometimes subtle cross-dependencies. Changes directly affecting a given system cannot be assumed to be neutral with regards to other systems; changes to (say) surface reflectivity, such as in the urban heat island effect, can in turn result in changes to rainfall patterns, influence the level of atmospheric ozone and particulate matter, and help determine the degree to which light from the Sun is absorbed.

Diversity is an argument against monocultures arising directly from and as an unintended consequence of human activity. Direct monocultures include commercial forest stands; unintended monocultures include the proliferation of aggressive invasive organisms (e.g., "weeds") after environmental shifts open up new niches. Monocultures make ecosystems less able to survive shocks.

Foresight recognizes that ecological and geophysical changes tend to be slow, in human terms; because of this, it's important when considering the implications of proposed actions to think in terms of the planet's pace, not just society's pace. An example would be the (as of now uncommon) recognition that global warming involves slow but relentless changes, such that quick shifts in human behavior will have no noticeable immediate effect.

Integration is an explicit counter to the "die-off" line of thinking that places the needs of human societies below all other systems on the planet. Not only does the "die-off" argument result in ecological disaster as desperate societies try to grab remaining resources, its logic leads to the argument that (a) since human society is inherently unsustainable, and (b) since the planet, given sufficient time, can recover from any environmental burden we place on it before we die, there's no reason to be cautious, and we should do as we like with no concern for the future. Seeing human societies as part of the planet's systems, and as worthy of preservation and protection as any other part, allows for a longer-term perspective.

Expansion of Options encompasses "sustainability," but is a larger concept. This means more than simply finding a sustainable balance of use and preservation; expansion of options means actively seeking behaviors that return more resources to the planet than they take, that emphasize renewal and reuse, and that provide a growing, diverse basis for future innovation.

Reversibility is an attempt to capture the idea that, where possible, we should bias towards those choices that allow for reconsideration if unanticipated and undesirable consequences arise. Reversibility will not always be an option — indeed, when matched with the Foresight principle, we may not recognize a problem until well after the option of reversal has passed. But when reversible options are available, they should be given special consideration.

LONG RUN, LONG LAG

All distant problems are not created equally.

By definition, distant (long-term) problems are those that show their real impact at some point in the not-near future; arbitrarily, we can say five or more years, but many of them won't have significant effects for decades. Our habit, and the institutions we've built, tend to look at long-term problems as more-or-less identical: Something big will happen later. For the most part, we simply wait until the long-term becomes the near-term before we act.

This practice can be effective for some distant problems: Let's call them "long-run problems." With a long-run problem, *a solution can be applied any time between now and when the problem manifests*; the "solution window," if you will, is open up to the moment of the problem. While the costs will vary, it's possible for a solution applied at any time to work. It doesn't hurt to plan ahead, but taking action now instead of waiting until the problem looms closer isn't necessarily the best strategy. Sometimes, the environment changes enough that the problem is moot; sometimes, a new solution (costing much less) becomes available. By and large, long-run problems can be addressed with common-sense solutions.

Here's a simple example of a long-run problem: You're driving a car in a straight line, and the map indicates a cliff in the distance. You can change direction now, or you can change direction as the

---

* *Original version of this piece published at Open the Future in October, 2008.*

cliff looms, and either way you avoid the cliff. If you know that there's a turn-off ahead, you may keep driving towards the cliff until you get to your preferred exit.

The practice of waiting until the long-term becomes the near-term is less effective, however, for the other kind of distant problem: Let's call them "long-lag problems." With long-lag problems, *there's a significant distance between cause and effect, for both the problem and any attempted solution.* The available time to head-off the problem doesn't stretch from now until when the problem manifests; the "solution window" may be considerably briefer. Such problems can be harder to comprehend, since the connection between cause and effect may be subtle, or the lag time simply too enormous. Common-sense answers won't likely work.

The nature of long-lag problems can be seen in steering oil tankers or stopping trains — a successful effort has to start long before when the turn or stop is needed. These are relatively short long-lag problems, however, operating on the scale of minutes. Real long-lag problems operate on the scale of years, even centuries, and we're not accustomed to thinking on that scale. Events that may have been set in motion years ago can simply seem like accidents or coincidences, or even assigned a false proximate trigger in order for them to "make sense."

Global warming is, for me, the canonical example of a long-lag problem, as geophysical systems don't operate on human cause-and-effect time frames. Because of atmospheric and ocean heat cycles (the "thermal inertia" I keep going on about), we're now facing the climate impacts of carbon pumped into the atmosphere decades ago. Similarly, if we were to stop emitting any greenhouse gases right this very second, we'd still see another two to three decades of warming, with all of the corresponding problems. If we're still three degrees below a climate disaster point, but have another two degrees of warming left because of thermal inertia *regardless of what we do*, we can't wait until we've increased to just below three degrees to act. If we do, we're hosed.

With long-lag problems, you simply can't wait until the problem is imminent before you act. You have to act long in advance in order to solve the problem. In other words, **the solution window closes long before the problem hits**.

We have a number of institutions, from government to religions to community organizations, with the potential to deal with long-run problems. We may not do well with them individually, but as a civilization, we've developed decent tools. However, we don't have many — perhaps any — institutions with the inherent potential to deal with long-*lag* problems. Moreover, too many people think all long-term problems are long-*run* problems.

(This argument emerged from a mailing list discussion of the Copenhagen Consensus. Smart people, with lots of good ideas, but clearly convinced that we can address global warming as a long-*run*, not long-*lag* problem.)

Sadly, recognizing the difference between long-run and long-lag problems simply isn't a common (or common-sense) way of thinking about the world. We evolved to engage in near-term foresight (and I mean that literally; look at the work of University of Washington neuroscientist William Calvin for details), and (as noted) we have developed institutions to engage in long-run foresight. Long-lag is a hard problem because it combines the insight requirements of long-run foresight (e.g., being able to make a reasonable projection for long-range issues) with the limited-knowledge-action requirements of near-term foresight (e.g., being able to act decisively and effectively before all information about a problem has been settled). Both are already difficult tasks; in combination, they can seem overwhelming.

A salient characteristic of long-lag problems is that they're often not amenable to brief, intense interactions as solutions. Dealing with such problems can take a long period, during which time it may be unclear whether the problem has been solved. Politically, this can be a dangerous time — the investment of money, time and expertise has already happened, but nothing yet can be shown for it.

This is why arguments that we should hold off even studying geoengineering are dangerous. Our responsibility to the future — our responsibility to ourselves — requires us to understand the consequences of choices we have yet to make. Waiting until geoengineering is our only option before thinking through the right course of action would be far too late.

OPEN SOURCE TERRAFORMING

Arguably, the geoengineering genie is already out of the bottle.

On February 9, 2007, Virgin Corporation honcho Richard Branson announced that he would give $25 million to the winner of the "Virgin Earth Challenge."

The Virgin Earth Challenge will award $25 million to the individual or group able to demonstrate a commercially viable design which will result in the net removal of anthropogenic, atmospheric greenhouse gases each year for at least ten years without countervailing harmful effects. This removal must have long term effects and contribute materially to the stability of the Earth's climate.

Environmentalist reaction to this announcement was cautiously optimistic, with most responses noting a comparison to the "X-Prize" for private space flight (although some observed that air travel, such as that provided by Virgin Airways, remains a big source of greenhouse gases). Much to my surprise, however, none of the major green blogs noted the most significant aspect of this competition:

This is explicitly a call for geoengineering projects.

The Virgin Earth Challenge isn't simply looking for better ways to reduce or eliminate new greenhouse gas emissions, it's looking for ways to remove existing CO2 and other greenhouse gases from the atmosphere — that's what "net removal" means.

---

* *Original version of this piece published at Open the Future in February, 2007.*

This competition seeks ways to make an active, substantial change to the Earth's geophysical systems. Richard Branson is underwriting terraforming, and given that the consensus new mainstream environmentalist position is to be solidly anti-geoengineering, the lack of reaction to what is essentially the "Terraforming Challenge" is a bit surprising.

But if we're already looking at geoengineering, and may potentially need to consider it as a necessary path to survival, how can we do it in a way that has the best chance to avoid making matters worse?

I've long argued that openness is the best way to ensure the safe development and deployment of transformative technologies like molecular nanotechnology, general machine intelligence, and radical human bio-enhancements. Geoengineering technologies should be added to this list. The reasons are clear: the more people who can examine and evaluate the geoengineering proposals, the greater the likelihood of finding subtle flaws or dangers, and the greater the pool of knowledge that can offer solutions.

As I put it in my 2003 essay for the final *Whole Earth* magazine (and the source of my blog's name), "Open the Future,"

Opening the books on emerging technologies, making the information about how they work widely available and easily accessible, in turn creates the possibility of a global defense against accidents or the inevitable depredations of a few. Openness speaks to our long traditions of democracy, free expression, and the scientific method, even as it harnesses one of the newest and best forces in our culture: the power of networks and the distributed-collaboration tools they evolve.

Broad access to... [transformative] tools and knowledge would help millions of people examine and analyze emerging information, nano- and biotechnologies [and geo-technologies], looking for errors and flaws that could lead to dangerous or unintended results. This concept has precedent: it already works in the world of software, with the "free software" or "open source" movement. A multitude of developers, each interested in making sure the software is as reliable and secure as possible, do a demonstrably better job at making hard-to-attack software than an office park's worth of programmers whose main concerns are market share, liability, and maintaining trade secrets.

[...] The more people participate, even in small ways, the better we get at building up our knowledge and defenses. And this openness has another, not insubstantial, benefit: transparency. It is far more difficult to obscure the implications of new technologies (or, conversely, to oversell their possibilities) when people around the world can read the plans.

The idea of opening transformative technologies is controversial. One argument often leveled against it is that it puts dangerous "knowledge-enabled" technologies into the hands of people who would abuse them. As noted earlier, the potential for non-state geoengineering efforts is very real, but while it's possible that an open-source approach would make those efforts somewhat easier, the complexity of geoengineering has more to do with successful implementation than as-yet unknown technologies. Another criticism of the open approach attacks it for undermining the market. But concerns about proprietary information and profit potential are hard to fathom with terraforming — there would be no plausible way to limit access to climate change remediation only to those who pay for it. Ultimately, the downsides of making potential geoengineering methods open are minor, while the benefits are massive.

Proprietary approaches and secret plans have a particular drawback: The value of the "many eyes" approach is enhanced if it isn't limited to after-the-fact analysis. Discovery of a flaw requiring a redesign is less costly — and less likely to be ignored — if it happens early in the development process.

Our best pathway to avoiding climate disaster remains the rapid reduction and elimination of anthropogenic greenhouse gases. But like it or not, we've entered the era of intentional geoengineering. The people who believe that it is a bad idea *need* to be part of the discussion about specific proposals, not simply sources of blanket condemnations. We need their insights and intelligence. The best way to make that happen, the best way to make sure that any terraforming effort leads to a global benefit, not harm, is to open the process of studying and developing geoengineering tools.

It may well be the best example yet seen of the importance of opening the future.

THE REVERSIBILITY PRINCIPLE

Two philosophies dominate the broad debates about the development of potentially-worldchanging technologies. The *Precautionary Principle* tells us that we should err on the side of caution when it comes to developments with uncertain or potentially negative repercussions, even when those developments have demonstrable benefits, too. The *Proactionary Principle*, conversely, tells us that we should err on the side of action in those same circumstances, unless the potential for harm can be clearly demonstrated and is clearly worse than the benefits of the action. In recent months, however, I've been thinking about a third approach. Not a middle-of-the-road compromise, but a useful alternative: the Reversibility Principle.

It's very much a work-in-progress, but read on to see what this could entail, and please feel free to add comments and critiques.

The Precautionary Principle, first articulated in 1988, argues that uncertainty should be a trigger for caution when it comes to technological advances. The most widely-accepted version of the principle comes from the Wingspread Statement:

When an activity raises threats of harm to human health or the environment, precautionary measures should be taken even if some cause and effect relationships are not fully established scientifically. In this context the proponent of an activity, rather than the public, should bear the burden of proof. The process of

---

* *Original version of this piece published at Worldchanging.com in March, 2006.*

applying the Precautionary Principle must be open, informed and democratic and must include potentially affected parties.

Transhumanism advocates Max More and Natasha Vita-More created the Proactionary Principle in 2004 as a direct counter to the Precautionary Principle. This concept argues that only probable and serious negative outcomes should be enough to block the development of potentially-useful technologies. The current version of the statement can be found on Max More's website:

People's freedom to innovate technologically is highly valuable, even critical, to humanity. This implies a range of responsibilities for those considering whether and how to develop, deploy, or restrict new technologies. Assess risks and opportunities using an objective, open, and comprehensive, yet simple decision process based on science rather than collective emotional reactions. Account for the costs of restrictions and lost opportunities as fully as direct effects. Favor measures that are proportionate to the probability and magnitude of impacts, and that have the highest payoff relative to their costs. Give a high priority to people's freedom to learn, innovate, and advance.

There's room for debate in each of these philosophies, of course. Many environmental activists subscribe to a version of the Precautionary Principle that focuses on taking responsibility for possible negative outcomes rather than simply avoiding any action that might lead to problems; Mike Treder, executive director of the Center for Responsible Nanotechnology, characterizes this as the "active" form of the Precautionary Principle. The

Proactionary Principle doesn't yet have multiple strongly-articulated versions, but the principle's authors have modified its wording in response to ongoing discussion; it's currently at version 1.2, although an earlier phrasing can be found in the Wikipedia article.

Critics of the Precautionary Principle claim that it focuses too much on worst-case scenarios, and gives insufficient weight to likely benefits of disputed technologies. Critics of the Proactionary Principle claim that it focuses too much on simple cause-and-effect logic, and ignores both complex results arising from interactions with other developments, and the potential for significant-but-not-inevitable problems. In my view, *both* of these arguments are largely correct.

We live in a world of rapid technological advances and tremendous global problems. Ideally, the first can help ameliorate the second; unfortunately, given the power of many of these advances, we run a strong risk that the first could make the second even worse. A binary "do it"/"don't do it" argument isn't well-suited to the degree of uncertainty that accompanies technological advances, nor the combinatorial, mutually-reinforcing aspects of global problems (such as climate disruption making conditions of poverty worse in the developing world, driving people towards survival strategies that further degrade the environment). I propose, instead, that we think not in terms of "caution" or "action," but in terms of *"reversibility."*

REVERSIBILITY

This is a draft articulation of the Reversibility Principle:

When considering the development or deployment of beneficial technologies with uncertain, but potentially significant, negative results, any decision should be made with a strong bias towards the ability to step back and reverse the decision should harmful outcomes become more likely. The determination of possible harmful results must be grounded in science but recognize the potential for people to use the technology in unintended ways, must include a consideration of benefits lost by choosing not to move forward with the technology, and must address the possibility of serious problems coming from the interaction of the new technology with existing systems and conditions. This consideration of reversibility should not cease upon the initial decision to go forward to hold back, but should be revisited as additional relevant information emerges.

Let's look at this in more detail.

"... *development or deployment...*" Ideally, the Reversibility approach would take hold in the early stages of the research and development process. The goal isn't necessarily to shut down research the moment potential problems are discovered, but to make certain to design the technology or process with reversibility in mind. We can assume that responsible technological development includes a desire to avoid harm; the Reversibility Principle would add to that a desire to include an "off switch" if harm is later identified.

"...technologies..." By this I mean any human-constructed tool, whether mechanical, biological or social.

"...uncertain, but potentially significant, negative results..." This encompasses two key issues: the negative results need not be guaranteed or inevitable; they should, however, be demonstrably serious. How "significant" is defined is likely to be a point of debate, but to start, I would look at the possibility of death, the difficulty of mitigation or amelioration, and the potential to make other, existing problems worse.

"...strong bias..." The potential for reversibility should be a critical issue as to whether to develop or deploy a technology, but shouldn't be the sole determinant. Other issues, such as the need to avert an even greater problem, will always come into play.

"...reverse the decision..." This is the cornerstone of the principle. Ideally, we would be able to recall the technology and undo the damage it has done should an unexpected negative result emerge. This will not necessarily be easy or even possible — but the difficulty of reversing the effects of an action arises, in part, from not taking reversibility into account during the design process.

"...grounded in science..." Misunderstandings, rumors or myths — even popular ones — should not be sufficient to cause a decision to hold off the development or deployment of useful technologies. At the same time, we must recognize that all science is contingent upon better information, and the inherent uncertainties of scientific study should not be cause to dismiss concerns as not "grounded in science."

"...*the potential for people to use the technology in unintended ways*..." Saying that something is safe if used correctly isn't the same as it being safe. If "the street finds its own uses for things," those uses will often be contrary to the manufacturer's instructions. In short, consideration of possible harmful results must include possible misuses and abuses of the technology.

"...*consideration of benefits lost*..." The strongest argument against the strict form of the Precautionary Principle is that it fails to account for the harm that could result from the lack of the new technology in the same way as it accounts for the harm that could result from its deployment. In a world of large-scale problems requiring innovative solutions, this is dangerously short-sighted. The potential for irreversible negative results coming from the use of the technology must be weighed against the irreversible negative results coming from its relinquishment.

"...*interaction of the new technology with existing systems and conditions*..." This will be the most difficult to measure part of the Reversibility Principle. New technologies do not exist in a vacuum. When deployed, they immediately become part of a larger technological ecosystem, and effects that, in isolation, may be essentially harmless can, in combination with other parts of the ecosystem, lead to serious problems. An example would be a biofuel plan that leads many food farmers to shift to fuel crops, at the expense of the availability of food for poverty-stricken regions.

"...*should not cease*..." Once a decision has been made to deploy or not to deploy a given technology, questions about the technology should not be forgotten. New discoveries and analysis may change the balance of issues around the decision, and what was once the right choice may in time become the wrong one. In short, the decision as to whether a technology is sufficiently reversible should itself be reversible.

WHY REVERSIBILITY?

Reversibility is something that would be useful for everyone to think about as they decide whether or not to adopt a particular tool or system, but the concept is particularly important for designers and planners.

From the design perspective, reversibility is something that should be part of the overall design process, much like sustainability. Just as it's easier to undertake a sustainable or "cradle-to-cradle" project by including the concept from the beginning, technology deployments are more likely to be reversible if the concept is inherent to the design, not simply an afterthought. For designers, then, the Reversibility Principle would advocate the question "how can we make this technology in a way that gives us the best ability to shut it off and undo any harm it might cause?" There may not be a perfect answer to the question, but it's almost inevitable that designs that take this issue into account will be more reversible than those that do not.

For planners, reversibility becomes an issue to take into account as technology development turns into deployment. By "planners," I mean anyone with responsibility for how a technological system gets into common use. For manufacturers, Reversibility Principle planning could be a hedge against lawsuits; for governments, Reversibility Principle planning could be a part of both economic and political strategy. If the reversibility concept were to take hold, I would imagine that insurance companies would be among its most strident advocates.

So how would the Reversibility Principle play out in practice?

One obvious candidate for reversibility analysis is biotechnology. A Precautionary approach says that we don't know the long-term effects of introducing genetically modified organisms (GMOs) into the ecosystem, as they are self-replicating technologies subject to evolutionary pressures; we should, therefore, avoid their deployment. Proactionary advocates argue that the benefits of the use of GMOs can be substantial, particularly in parts of the world that (for political or environmental reasons) are unable to grow enough food for local populations; we should, therefore, encourage their development. As before, both of these positions are, in my view, more or less correct.

A Reversibility Principle approach to biotechnology in general would argue that GMOs should be engineered in a way to make it possible to remove them from the environment if unexpected or low-probability problems emerge. Issues of human consumption of GMOs would be handled on a case-by-case basis, with a bias towards holding off on products that demonstrate a possibility of serious or irreversible problems.

Another candidate for the reversibility approach is the response to global warming. The Precautionary Principle *and* the Proactionary Principle could each be use to justify both rapid action to reduce carbon and a "wait for better methods" approach.

From a Reversibility Principle perspective, however, the choice is clear. The potential problems arising from immediate action to cut carbon emissions are largely economic, and while in the worst case scenario they are serious, they are more easily mitigated than those that would come from a slow response, which in even a moderate-case scenario would harm hundreds of millions of people in irreversible ways.

The Reversibility Principle would clearly apply in the case of geoengineering. It's likely that, should we be forced to consider such global-scale engineering to respond to climate disaster, few of the options will be reversible. The question then becomes which option — including the option of doing nothing — would in the worst reasonable scenarios result in the least amount of death and destruction, and which would give us the greatest opportunity for gradual mitigation of harm. Underlying the choices will be the need to make the ways the options as reversible *as possible*, even if full reversibility isn't plausible.

There are two major questions that come to mind about the Reversibility Principle. To be blunt, the first is whether "reversibility" is even possible. From a purely physical perspective, it's not; even the act of stepping back and brushing over one's footprints still shifts the sand. But there's a difference between being unable to return the world exactly to how it once was and being unable to avoid inevitable disasters. Some of the difference arises from how soon we decide that a choice needs to be reversed; even gradual changes can become irreversible if given enough time to accumulate.

We should see "reversibility," then, not as an attempt to go back to precisely how the world once looked, but as an attempt to eliminate further harm by its source, and to ameliorate the harm that has occurred.

But the bigger issue for the Reversibility Principle perspective is just how readily we can predict the various possible outcomes, both good and bad. The quick answer is we can't fully, but that hasn't stopped us from planning for the future before; we often need to act in situations of limited information. This doesn't mean our choices must be ill-informed.

This is a situation where Scenario Planning methodology could be of value. The scenario approach intentionally avoids coming up with a single "most likely" future. Instead, scenario planners come up with multiple contingent futures, with none of them meant to be a prediction. Rather, the collection of scenarios function as environments in which to test plans — strategic wind tunnels, if you will. In Reversibility analysis, planners would come up with multiple contingent futures in which to think about outcomes if the given technology is or is not deployed.

There's also the possibility of increasingly sophisticated models and simulations. I have enough experience in the use of computer models for political and social analysis to know that simulations should stick to physical systems, but it may be possible in time to develop decision-making aids using computer models that help human decision-makers to better understand both the physical and social dynamics at work. In situations where harmful outcomes are highly contingent but potentially very serious, good simulations could help answer the "what happens if..." questions in ways that can better be applied to questions of reversibility.

REVERSIBILITY AND THE OPEN FUTURE

A cornerstone of the open future concept is that we should be striving towards a world that maximizes our flexibility in response to challenges. We will never have perfectly free choices when problems arise, but we are more likely to come up with good solutions under less-constrained conditions than we would if we were limited to a handful of options. The choice to pull back and say "let's try something different" is an option that we should strive to maintain.

Ultimately, the Reversibility Principle should be a heuristic, a prism through which we look at the world and make our decisions. We may not always choose the path with the simplest way back — it may not always be the right choice — but it would encourage us to consider the issue for all of our options. Asking ourselves, "if we do this, how readily can it be undone if we discover problems?" forces us think in terms of more than immediate gratification, and to consider how the choice connects to other choices we and the people around us have made and will make. In the end, it may even be a good first-order approximation of wisdom.

HARD CHOICES

UNDERSTANDING OUR OPTIONS

Depopulation is not a global warming strategy.

Here's what leads me to that (seemingly obvious, but apparently not) observation.

We know these to be true:

Feedback effects ranging from methane released from melting permafrost to carbon emissions from decaying remnants of forests devoured by pine beetles will boost greenhouse gases faster than natural compensation mechanisms can handle.

The accumulation of non-linear drivers can lead to "tipping point" events causing functionally irreversible changes to geophysical systems, such as massive sea-level increases. Some of these can have feedback effects of their own, such as the elimination of ice caps reducing global albedo, thereby accelerating heating.

Because of the long, slow nature of carbon cycles, no matter what we do, we are committed to warming the planet for at least two to three decades beyond whatever point we finally stop adding to greenhouse gases.

We also know these to be likely:

The economic, environmental and social benefits accruing to early adopters of cleaner infrastructure and behavior can serve as a catalyst for faster adoption by lagging actors. In short, the first ones in demonstrate that the water's fine.

---

* *Original version published at Open the Future in April, 2008.*

Many of the cleaner technologies, infrastructure and behavior have ancillary benefits, from quality-of-life to political rebalancing, that can accelerate their adoption.

Continued technological innovations could allow for faster mitigation of greenhouse gases, even potentially allow for the uptake of atmospheric carbon, accelerating the natural cycle of carbon from the atmosphere.

So: we have a set of demoralizing forces at play, countered by a set of encouraging possibilities. What is the common element that would allow those possibilities to play out? **Time.**

Time is what we need.

Time is what we may not have.

Climate and environmental sciences remain imperfect, but few of the improvements in our understanding have reduced the sense of urgency surrounding global climate disruption. On the contrary, much of the enhanced analysis has increased scientists' level of worry. Richard Clarke once famously described a subset of international security analysts running around Washington DC in 2000 and 2001 with their "hair on fire," trying to alert policy-makers to the potential for a terrorist attack in the US. Today, it's the geophysical scientists with their hair on fire, sounding increasingly desperate and shrill about delays in responding to climate meltdown. And they have good cause for alarm: even an enlightened transition away from business-as-usual energy, transportation and social systems may not happen fast enough to avoid catastrophe; certainly, the slow, mulish pattern we've seen up to the present won't.

If it all comes down to time, we have two choices: move faster, or get more time.

Moving faster is the approach preferred by nearly everyone making a study of climate and environmental changes. We know what we need to do, we know roughly what it will cost and how long it will take, and we know ways to make it happen to all of our benefit. Unfortunately, we apparently have bigger priorities at the moment, and will get to this climate thing when it really starts to make some noise (by which time, it will be far too late). It seems we're just not that good at thinking in terms of lagging cause-and-effect, and the need for long-term thinking.

We could get lucky; positive feedbacks and "the water's fine" demonstrations may allow us to move faster.

We could also get "lucky" in a not-so-lucky way: a clarity-inducing global disaster could trigger the necessary economic and political shifts without pushing us over the edge. Arguably, a series of even moderate natural disasters that could be convincingly tied to global warming (convincing at the political level, even if scientists remain cautious) might serve as a goad to get recalcitrant actors to move faster or suffer political harm. It wouldn't be so lucky for the thousands or millions of people suffering from these "clarity-inducing" disasters, of course, or for the thousands or millions who would suffer from subsequent disasters happening while we get ourselves in gear.

Getting more time means slowing down the greenhouse gas-heat-feedback cycle, and that means geoengineering. We're still in the research phase with this; it's not something to try today. The most important task for current geoengineering work is to identify the approaches that might look attractive at first, but have devastating results — we need to know what we should avoid even if desperate.

Make no mistake: I am *not* arguing that geoengineering, should it be tried, would be a replacement for making the economic, social, and technological changes needed to eliminate anthropogenic greenhouse gases. It would *only* be a way of giving us more time to make those changes. It's not an either-or situation; geo is a last-ditch prop for making sure that we can do what needs to be done.

Claims that we shouldn't even talk about geoengineering, or give it any kind of meaningful research funding, while we're trying to get people to move faster smacks of Condoleezza Rice's infamous statement regarding contingency planning and the Iraq war:

"It's bad policy to speculate on what you'll do if a plan fails when you're trying to make a plan work."

No. Wrong. Sorry. The only rational, resilient, *ethical* approach is to prepare to deal with failure of one's preferred strategy *before* that failure occurs. I don't want us to have to engage in geoengineering. I want us to stop being such idiots and start to make real changes to our societies, our infrastructure, our lives.

But I also know that we're getting awfully close to the point of being too late for those changes to have a meaningful impact.

And if we're too late, millions, perhaps billions, of people will die. I will not accept the loss of so many lives as the only alternative to political leaders in the US and China getting their acts together. **Depopulation is not a global warming strategy.** It's a horrific, tragic result of the failure of strategy, the failure of imagination, and the failure of our capacity to fight to the last breath for our future.

SAVING THE PLANET, SAVING OURSELVES

The grand myth of environmentalism is that it's all about saving the Earth.

It's not. The Earth will be just fine. Environmentalism is all about saving ourselves.

That may seem a bit counter-intuitive; after all, the Earth is certainly central to the rhetoric, the memetics of environmentalism. Most environmental discussions focus on ecological dynamics, with references to human beings typically limited to enumerations of the various insults we've visited upon the planet. Given the degree of culpability we bear for the current state of the planet, this is entirely appropriate.

But the rhetorical focus of environmentalism on the planet obscures the fact that what human beings have done to the Earth pales in comparison to past disasters hitting our world, from massive asteroid strikes to super-volcano eruptions killing off 90+ % of the Earth's species. And in every case, the Earth has recovered, and life has once again flourished.

We sometimes make the conceptual mistake of thinking that the way the Earth's ecosystem is today is the way it will forever be, that we've somehow reached an ecological end-state. But even in an eco-conscious world, or one devoid of humans entirely, natural processes from evolution to geophysical and solar cycles would continue. The Earth's been at this for a *long* time, literally

* *Original version published at Open the Future in April, 2008.*

billions of years; from a planetary perspective, a quadrupling of atmospheric carbon lasting 10,000 years (for example) is little more than a passing blip.

The fact of the matter is that, no matter how much greenhouse gas we pump into the atmosphere or how many toxins we dump into the soil and oceans, given enough time the Earth — and its ecological systems — will recover.

But human civilization is far more fragile. Human civilization could not withstand and recover from the same kinds of assaults the planet itself has shrugged off in eons past. We remain entirely dependent upon myriad Earth services and systems, from topsoil and clean water to carbon cycles and biodiversity. Activities that undermine those critical services and systems quite literally threaten the survival of human civilization. The fundamental resilience of the Earth's geophysical systems simply means that, when we ignore our effects on the planet, we're simply making ourselves disposable, just another passing blip in the planet's long history.

In trying to minimize the harmful impacts of human activities upon the global ecosystem, environmentalism supports the continued healthy existence of humankind.

To me, this too is entirely appropriate. Despite its many flaws, I'm a big fan of human civilization. I marvel at our capacity to organize matter and information, at our ability to learn from mistakes and pass that learning down to subsequent generations. Civilization — writing, cities, trade, the whole lot of it — makes us unique on this planet and, as far as we can tell so far, in our part of the universe. Destroying that through malice or negligence is the worst form of crime, and the height of tragedy.

Part of a focus upon civilization, however, is the recognition that we do not exist in isolation, that we are dependent upon an enormous variety of complex systems. As a result, our continued existence requires the continued success of those systems. In order to save ourselves, we have to minimize actions which damage and disrupt the environment.

Like any social movement, environmentalists argue over tactics and goals, and some eco-activists will disagree with my characterization of the purpose of environmentalism. But the reality is that — at least with *current* technologies — there's nothing that we can do to truly put the planetary biosphere at existential risk. It *will* recover from what we now do, albeit in a different form than today. But what we can do is so violate the integrity of the planet's ecosystem that the Earth can no longer support *us*.

Critics of environmentalism often claim that eco-activists hate humans, that we value the Earth more than we value ourselves. With very few exceptions, nothing could be further from the truth.

Environmentalism is fundamentally about making sure that human beings, and human civilization, can continue to thrive on our home planet for centuries, millennia to come. Environmentalism, in its demands for respect for nature, ultimately demands that we respect ourselves.

Sources and linked references can be found at the original web entries for these articles:

INTRODUCTION: SOLVING THE CLIMATE CRISIS
http://www.openthefuture.com/2007/10/solving_the_climate_crisis.html

THE GEOENGINEERING OPTION
http://futurismic.com/2006/10/01/new-column-jamais-cascio-on-the-geoengineering-option/

GEOENGINEERING AWARENESS
http://www.openthefuture.com/2008/07/more_geoengineering_coverage.html

COMPARING GEOENGINEERING TECHNIQUES
http://www.openthefuture.com/2009/01/new_geoengineering_study_part.html

THE QUESTION OF METHANE
http://www.worldchanging.com/archives//003283.html

GLOBAL CLIMATE AND GLOBAL POWER
http://www.openthefuture.com/2008/12/global_climate_and_global_powe.html

THE POLITICS OF GEOENGINEERING
http://www.openthefuture.com/2007/10/the_politics_of_geoengineering.html

TERRAFORMING WAR
http://www.foreignpolicy.com/story/cms.php?story_id=4146

GEOETHICAL PRINCIPLES
http://www.openthefuture.com/2007/01/otf_core_geoethical_principles.html

LONG-RUN, LONG-LAG
http://www.openthefuture.com/2008/10/long-run_vs_long-lag.html

OPEN-SOURCE TERRAFORMING
http://www.openthefuture.com/2007/02/open_source_terraforming.html

THE REVERSIBILITY PRINCIPLE
http://www.worldchanging.com/archives//004174.html

UNDERSTANDING OUR OPTIONS
http://www.openthefuture.com/2008/04/feedback_tipping_points_and_ha.html

SAVING THE PLANET, SAVING OURSELVES
http://www.openthefuture.com/2008/04/the_earth_will_be_just_fine_th.html

ABOUT THE AUTHOR

Jamais Cascio writes about the intersection of emerging technologies, environmental dilemmas, and cultural transformation, specializing in the design and creation of plausible scenarios of the future. His work focuses on the importance of long-term, systemic thinking, emphasizing the power of openness, transparency and flexibility as catalysts for building a more resilient society.

Cascio's work appears in publications as diverse as *Metropolis*, the *Atlantic, Technology Review*, and *ForeignPolicy.com*. He was featured in National Geographic Television's **Six Degrees**, its 2008 documentary on the effects of global warming, and on History Channel's **Science Impossible**, its 2009 series on emerging technologies. Cascio has spoken about future possibilities around the world, from Singapore to Texas.

Recent projects of note include scenario design for the 2008 "massively multiplayer forecasting game," **Superstruct**, for the Institute for the Future, creating the world of 2019 inhabited by the game's thousands of players, and co-writing the 2007 **Metaverse Roadmap Overview**, a cross-industry examination of the next decade's evolution of online technologies.

In 2007, his work on calculating the carbon footprint of cheeseburgers went viral, appearing in dozens of newspapers and magazines, multiple radio programs, hundreds of websites, and even as part of a museum exhibit.

Cascio has worked in the field of scenario development for over a decade, and is currently a Research Affiliate at the **Institute for the Future**. After several years as technology specialist at scenario planning pioneer **Global Business Network**, he went on to craft a wide array of scenarios on topics including energy, nuclear proliferation, and sustainable development. Cascio serves as the Director of Impacts Analysis for the **Center for Responsible Nanotechnology**, and is a Fellow at the **Institute for Ethics and Emerging Technologies**.

In 2003, he co-founded and was the lead writer at **WorldChanging.com**, the award-winning website dedicated to finding and calling attention to models, tools and ideas for building a "bright green" future. In March, 2006, he left Worldchanging and started **Open the Future** (http://www.openthefuture.com) as his online home.

Cascio lives outside of San Francisco, California, with his wife, two cats, four Macs, and the inevitable hybrid cars.

Send Jamais email at cascio@openthefuture.com.

www.ingramcontent.com/pod-product-compliance
Lightning Source LLC
Chambersburg PA
CBHW022021170526
45157CB00003B/1311